THE RESILIENCE OF ECOSYSTEMS

RENÉ DUBOS

THE RESILIENCE OF ECOSYSTEMS

AN ECOLOGICAL VIEW OF ENVIRONMENTAL RESTORATION

COLORADO ASSOCIATED UNIVERSITY PRESS

Colorado Associated University Press
Boulder, Colorado 80302
International Standard Book Number 0-87081-107-X
Printed in the United States of America
Designed by Bruce Campbell

This is the first of the Gustavson Lectures to be delivered annually at the University of Colorado. Fourteen more will follow.

THE REUBEN G. GUSTAVSON MEMORIAL LECTURES were established in 1975 in recognition of Dr. Gustavson's contributions to education and research. Sponsored by Resources for the Future, Inc., of Washington, D.C., and four universities he served, the lectures will be given annually for fifteen years at the University of Arizona, the University of Chicago, the University of Colorado, and the University of Nebraska. Contributions to the lecture fund were made by Resources For the Future, Laurance S. Rockefeller, William S. Paley, and by former students and colleagues at each of the sponsoring institutions. The lectures deal with Science and Society, with emphasis on questions related to resources and the environment.

René Dubos's lecture, the first of the series, was delivered at the University of Colorado on January 24, 1977.

HUMAN ACTIVITIES AND ENVIRONMENTAL QUALITY

Humans have spread over the whole surface of the Earth, but paradoxically they are biologically out of place in most of the natural environments in which they have settled. The reason for this paradox is that the genetic constitution of the human species has not changed significantly since the late Stone Age, whereas the immense majority of humans now live in regions very different from those where our species acquired its fundamental biological characteristics. Since humans now occupy natural environments to which they are not biologically adapted, they must transform them to fit their own biological needs. They could not survive long even in the temperate zone if it were not for the humanized habitats they have created by using fire, developing agriculture, and manufacturing a great diversity of artifacts. Human life implies the humanization of the Earth.

I am of two different minds when I consider the consequences of the transformations of the Earth that have resulted from human activities. On the one hand, I am frightened by the propensity of our species to desecrate the Earth, to pollute its environments, to waste its resources. I shall come back later to the problems of environmental degradation in the modern world. On the other hand, I believe that many human interventions into Nature have been beneficial not only to humankind but also to the Earth.

For several decades, discussions concerning the environment have emphasized almost exclusively the damage done to ecosystems by the human presence; these discussions have tended to assume furthermore that the damage is largely irreversible. For example, when the United Nations Conference on the Human Environment convened in Stockholm in 1972, some of its main themes could be summarized by phrases such as: rivers and lakes are dying; deserts are on the march; natural resources are being depleted; pollution is making the Earth unsuited to life. Many people have indeed come to believe that humans cause damage to the Earth whenever they disturb the natural order of things.

In his celebrated essay "A Sand County Almanac," Aldo Leopold asserted that "A thing is right when it tends to preserve the integrity, stability and beauty of the biotic community. It is wrong when it tends otherwise."[1] This conservative view of the relationship between humans and their environment is now regarded as the great commandment of conservation philosophy: Leopold referred to it as a "land ethic." In my judgment, however, it presents an incomplete picture of the interplay between humankind and the Earth.

The statement that we should "preserve the integrity . . . of the biotic community" is of course sound, but it assumes that natural processes have created the best possible biotic community and indeed the only viable ecosystem. There is no evidence for this assumption. Examples will be quoted later to illustrate that some of the world's most diversified, beautiful and productive ecosystems have been created by human activities which have profoundly transformed natural environments. Many potentialities of the Earth remain unexpressed in the state of wilderness and have to be brought out by human imagination, knowledge and toil. While Nature is obviously the origin of all raw materials and all forms of energy, it does not always use them to the best possible advantage. We humans can use natural resources to create new ecosystems which are ecologically sound, economically productive, and esthetically rewarding.

Increasingly, however, human activities are causing environmental damage all over the earth. Deforestation, erosion, and salination are lowering agricultural productivity in many places. Chemical pollutants are spoiling the quality of air, water, and other components of ecosystems with disastrous effects on human beings and other forms of life. The widespread awareness of the dangers inherent in social and technological innovations is generating intense concern for the long-range human and environmental effects of these innovations, as indicated by the legal necessity to work out statements of environmental impact for each important new enterprise; but many persons fear that awareness of environmental degradation has come too late because much of the damage already done to ecosystems is irreversible. In my opinion, this pessimism is unjustified because ecosystems have enormous powers of recovery from traumatic experiences.[2]

Ecosystems possess several mechanisms for self-healing. Some of these are analogous to the homeostatic mechanisms of animal life; they enable ecosystems to overcome the effect of disturbances simply by reestablishing progressively the original state of ecological equilibrium. More frequently, however, ecosystems undergo adaptive changes of a creative nature that transcend the mere correction of the damage; the ultimate result is then the activation of certain potentialities of the ecosystem that had not been expressed before the disturbance.

There are numerous examples of such environmental recovery occurring either through simple homeostatic response, or through adaptive changes of a creative nature. I shall describe a few of them, selected to illustrate environmental problems in different climatic zones and to represent different technical approaches to environmental improvement.

HOMEOSTATIC RECOVERY OF NATURAL ECOSYSTEMS

Anyone who has established a home on abandoned farmland in the temperate zone is painfully aware of Nature's ability to restore the original forest vegetation which once covered much of the Earth. This has been my own experience in the Hudson Highlands forty miles north of New York City, where I have to struggle endlessly against the natural processes of plant succession that created the primeval forest in that part of the world. Throughout the temperate zone, the forest soon regains its former prominence unless kept under control by uninterrupted human effort.

A recent bulletin from the University of Rhode Island Experimental Station provides a typical illustration of the restorative ability of Nature in the temperate zone.[3] As far back as two centuries ago, 70 percent of the land area in the state of Rhode Island had been cleared of the deciduous forest that once covered it almost completely. The primeval forest had been transformed into agricultural land by the original white settlers. During the late nineteenth century, however, the less productive farms were abandoned and trees returned so rapidly that less than 30 percent of the whole state remains cleared today. Nature provided the mechanisms for a spontaneous step-by-step restoration of the original ecosystem. Similarly, the forest is reoccupying abandoned farmlands in many other areas of the eastern United States.[4]

In Western Europe the process of deforestation began during the Neolithic period and probably reached its peak a century ago, but marginal agricultural land is now being abandoned for economic reasons. Wherever this happens, the brush and trees take over. The original type of forest progressively becomes reestablished even on land that had been cleared and used for crops or pastures for several thousand years!

Although trees are the most obvious manifestations of ecologic recovery in the temperate zone, they are not the only ones. Animals also reestablish themselves as soon as they have a chance. Deer multiply to a nuisance level in all areas where land management provides them with an adequate supply of food; wild turkeys have once more been sighted in all counties of New York State; coyotes and even wolves are on the increase in those parts of the northeastern states which are reacquiring some wilderness characteristics.[5]

The reintroduction of beavers in Sweden provides a picturesque illustration of Nature's ability to reestablish the natural order after it had been destroyed by human intervention. In 1871 the last of the original population of beavers was shot in Sweden. By that time this species had disappeared from most of its former European habitats. However, when a few beavers from the surviving Norwegian stock were reintroduced in Sweden between 1922 and 1939, they rapidly multiplied to such an extent that they caused extensive damage to forest and arable land. There are once more loud demands in Sweden for an open season against beavers and even for their complete eradication.[6]

Admittedly, the restorative processes of Nature are much less effective in the other parts of the world than they are in the temperate zone. Despite their massive grandeur and seemingly stark immutability, the Himalayas, the Andes and the East African mountains are among the most fragile ecosystems on earth. Their steep slopes deteriorate rapidly and perhaps in an irreversible manner when erosion follows overuse in the form of excessive woodcutting, grazing or cropping. Similarly, semi-desertic areas and tropical humid forests also are extremely susceptible to environmental insults. All over the earth, deserts are on the march.[7]

Yet even the most fragile ecosystems can recover under special circumstances. Recovery can take place, for example, through the agency of species that reach the damaged ecosystem either by accidental transportation or by active migration. In 1883, the Krakatoa island in the Malay peninsula was partly destroyed by a tremendous volcanic eruption that killed all its forms of life.[8] Experts have estimated that the explosion had the violence of a million hydrogen bombs. The seismic wave it generated reached up to 135 feet above sea level, destroying seaside villages in Java, Sumatra, and neighboring islands. Ash and gases rose 50 miles into the sky, blocking out sunlight over a 150-mile radius. Vast quantities of pumice hurtled through the air, defoliating trees and clogging harbors. When the eruption ended, what was left of the island was covered by a thick layer of lava and was completely lifeless.

The wind and the sea currents, however, soon brought back some animals and plants, and life once more took hold on the lava. More than 30 species of plants were recognized as early as 1886. By 1920, there were some 300 plants species and 600 animal species including birds, bats, lizards, crocodiles, pythons and of course rats. Today, less than a century after the great eruption, the plant communities on Krakatoa are approaching the composition of the climax forest in the rest of the Malay Archipelago. Many examples of such resurgence of life have been observed under other conditions. Even Bikini and Enivetok, pulverized and irradiated by 59 nuclear blasts between 1946 and 1958, are said to be reacquiring a semi-normal biota, despite the destruction of their top soil.

The most recent illustration of Nature's biological power is the rapid establishment of living things on Surtsey, the new island created by a submarine volcanic eruption on November 14, 1963, off the coast of Iceland. Within less than ten years after its emergence, Surtsey had acquired from the neighboring islands and from Iceland itself a complex biota which makes it an almost typical member of the Icelandic ecosphere.[9]

The introduction of biota from a foreign source is not always needed for the recovery of a fragile ecosystem. Plants or their seeds can persist in a dormant state for long periods of time and prosper again as soon as conditions are favorable for their development.

The Wadi Rishrash region of the Eastern Desert in Egypt was shut off to grazing in the 1920s. Within a few years, the vegetation was so dense that it looked like that of an irrigated oasis; desert animals took refuge in it during the breeding season. The region appeared almost out of place in its barren surroundings.[10]

Similarly, in Greece and in the African Sahel, protection against grazing by cattle, goats, and rabbits is sufficient to enable a diversified vegetation to become spontaneously reestablished. Even trees reappear and grow rapidly in areas that had long been almost desertic. In a particular Sahelian ranch of one-quarter million acres, the use of barbed wire to prevent uncontrolled grazing and the policy to let cattle graze in each area of the ranch only once in five years, enabled the land to move spontaneously from the state of desert to that of pasture. Deserts are indeed on the march over much of the Earth, but they can be made to retreat.[11]

One of the most spectacular recoveries of a severely damaged ecosystem occurred recently in west Texas, near the city of San Angelo at the confluence of the three Concho rivers. The phenomenon seems to have begun in the valley of Rocky Creek, a stream that dried up thirty years ago, but that now flows once more so abundantly and regularly that it makes an important contribution to the municipal water supply of San Angelo. At the turn of the century, Rocky Creek was a never-failing clear stream that ran through a valley of tall grasses, dotted occasionally here and there by mesquite bush or some other form of brush. Fish and waterfowl were abundant in the stream; deer and smaller game sought refuge from summer heat under the trees of its banks. Throughout the early decades of the twentieth century, however, mesquite and other brush increasingly invaded the floor of the valley and hillsides. Rocky Creek became progressively narrower and shallower and eventually ceased to flow during the drought of the 1930s. Fish, waterfowl, deer, and other game virtually disappeared.

Although the drought contributed of course to the improverishment of Rocky Creek, much of the damage certainly resulted from changes in land use. Before the white settlements, herds of buffalo and other grazing animals periodically migrated down from the plateaux. They were so numerous that they left in their wake a hoof-scarred land almost denuded of grass, but the damage they caused was only transient because they stayed for a relatively short time. As they did not return to the same place for a year, or perhaps for several years, grass had time to grow back. The situation changed when the white ranchers made it a practice to enclose their cattle in barbed wire fences, so that the animals remained concentrated on the same area year in and year out. As a result of overgrazing, the better types of native grass progressively disappeared and were replaced by mesquite and other brush. The deep-growing roots of these plants sapped the underground water that used to permit the growth of the more desirable varieties of grasses, and that also found outlets into creeks and rivers.

Around the middle of the twentieth century a few ranchers began to change their grazing practices in order to protect their cattle against the drought, and to improve the quality of the range. They reduced livestock numbers to a level of grazing that enabled the tall grasses to return; simultaneously they started a program of brush destruction by chemical means and with herbicides. The result was beyond expectation. In 1964, a half-forgotten spring began to flow for the first time in some thirty years. Its flow progressively increased and soon Rocky Creek also came back to life. As more and more brush was eliminated, new seeps and springs began to flow in the valley. Many of them continued to yield clear water even during a hot, dry summer. Rocky Creek has flowed the year around, since the late 1960s, all the way to the Middle Concho. Fish, waterfowl, deer, and other game are once more part of the scenery. Other programs of brush control now conducted by the University of Texas in collaboration with ranchers show that, as was done in the Rocky Creek area, it is possible also in other areas to bring back to life numerous seeps and springs that had been dried up for several decades.[12]

The recovery of lakes and waterways that had been polluted by industrial and domestic effluents constitutes another manifestation of the restorative ability of Nature. In several parts of the world, ecological damage caused by water pollution has been completely or at least partially corrected, not by treating the polluted ecosystem but simply by interrupting the pollution process and letting natural forces eliminate the accumulated pollutants. The phenomenal results achieved for the Thames in London, the Willamette River in Portland, Oregon, Lake Washington in Seattle, and Jamaica Bay in New York City are but a few among the many examples of improvement in water quality achieved by anti-pollution measures during the past decade.[13] Naturally it would be impossible to present here the details of these encouraging achievements, but a few facts concerning Jamaica Bay will illustrate the type of results that can be obtained even under the least promising conditions.

Jamaica Bay is a large Atlantic bay adjacent to the J.F. Kennedy Airport within the limits of New York City. Until a few decades ago, it had an important shellfish industry and harbored immense numbers of migrating waterfowl—in the hundreds of thousands—during the spring and fall.

The Bay suffered extensive ecological damage because of its proximity to the large populations of the New York City area and also to the airport. Sand was dredged from its bottom; the marshes on its periphery were filled with garbage; its water was polluted by the discharges from more than 1600 sewers.

During the past two decades, however, attempts have been made to save the Bay. Water pollution control facilities have been established; the dumping of garbage has been discontinued; grasses and shrubs have been planted on the existing garbage islands. As a result, shellfish, finfish, and birds are once more abundant. The center of the Bay has become a wildlife refuge. The bird population is remarkable not only for its abundance, but also for its diversity and for the presence of a few rather unusual species. There are large numbers of wading shore birds such as sand pipers, dowitchers, green herons. The time of migration brings wave upon wave of scaup and brandt, mallards and canvas backs, Canadian geese and teal. The glossy ibis also has come back, as well as the snowy egret, a bird that was almost extinct in the 1920s.

The return of the glossy ibis in Jamaica Bay, of the wild turkey and the peregrine falcon in their old habitats of the eastern United States, of the salmon in the Willamette River and more recently in the Thames, illustrates that under the right conditions some of the original ecological order reestablishes itself spontaneously, once the disturbing influences have been discontinued.

Similar phenomena of ecological recovery have been observed in many other parts of the world, particularly in North America and Europe. It is probable therefore that environmental degradation can be interrupted in most cases and that the rate of improvement can often be more rapid and perhaps also less expensive than commonly believed. Following a human intervention of the right kind, Nature will often take over and heal itself. What is needed is not esoteric knowledge and technologies but simply good management and social will.

THE EVOLUTION OF NATURAL ECOSYSTEMS

Today's natural ecosystems are profoundly different from what they were during earlier geological times, and even different from what they were a few thousand years ago when the earth's climate had become essentially what it is now. Ecosystems constantly evolve under the influence of countless physico-chemical forces that are poorly understood and that are much more powerful than those unleashed by human activities. Long before the human presence, for example, stupendous dust storms had repeatedly occurred on Earth, as they have recently been shown to occur on the planet Mars. Climatic changes, volcanic eruptions, earthquakes, hurricanes, fires of natural origin, and the activities of animals and plants have also always played their part in the formation of landscapes.

During the first half of this century, the evolution of natural ecosystems was commonly regarded as corresponding to an orderly succession of plant and animal species following each other in a fairly well-defined order; according to this view, the end result was a climax population that remained fairly stable until it was upset by some major disturbance. In reality, however, such evolution is far more complex and more interesting than suggested by this petrified picture of succession and climax. Even without conscious human intervention, many random events influence the evolution of ecosystems in ways that can be hardly predicted.

Fire, for example, was considered for a long time to have only destructive effects on natural ecosystems, but there is now convincing evidence that certain species of plants depend on it for their continued existence in Nature. In most ecosystems, repeated small fires prevent the accumulation of fuels on the ground and thus minimize the dangers of catastrophic wildfires, but the extent of this effect is uncertain since the occurrence of fires is unpredictable under natural conditions. Probably much more important is the fact that fire plays a vital role in the development of certain plant species because it is essential to such processes as seedbed preparation, nutrient recycling, and plant succession. Indeed, fire is now believed to play such an important role under natural conditions that the National Park Service has recently adopted the policy to let fires of natural origin run their course almost unchecked in certain wilderness areas. Controlled fires are even started wherever necessary for the maintenance of a species or the health of an ecosystem.[14]

Fires have affected the evolution of ecosystems in several parts of the globe by preventing the spread of the forest or destroying it where it existed. This is particularly true for the prairie country of the North American Great Plains. Fires set by pre-agricultural Indians to facilitate the hunt of large animals retarded or prevented altogether the growth of trees and shrubs. They also released mineral nutrients from organic debris and made them available in great abundance to annual grasses.[15]

In the arid cattle-grazing lands of southern Arizona fire prevention programs have had the result of allowing the establishment of shrubs and trees such as mesquite and cholla, at the expense of grass. Once this new vegetation was established, moreover, it utilized and reduced the moisture supply, thus preventing the return of the grass even when grazing was discontinued.

An essential element in the growth of the prairie ecosystem was its animal population. Until the turn of the century, immense herds of buffalo trampled open spaces in the grass and at the same time richly manured the soil. In the words of a nineteenth-century observer, the buffalo "press down the soil to a depth of 3-4 feet. . . . All the old trees have their roots bare of soil to that depth." The spaces opened in the grass by the buffalo were utilized by smaller animals, such as prairie dogs, which themselves not only supported such predators as the black-footed ferret but also turned over enormous quantities of earth by their incessant burrowing activities.[16]

Whatever the factors involved in the evolution of the prairie vegetation, it is certain that the final result was a balanced system with luxurious and tall grasses, dozens and dozens of species of wildflowers, and a black sod more than ten feet deep in certain places. So many random events were involved in the emergence of the American prairie that it would probably be impossible to recreate this ecosystem today even if all its plants were known. The flora could not be reestablished in its original state without the participation of all the ancient fauna and other natural forces, including the trampling of the soil by immense herds of buffalo.

There are also European ecosystems which acquired their characteristics through the effect of animals on the soil. During the middle of the fourteenth century, for example, large areas of arable land in England were abandoned as a result of the plague epidemic (Black Death) which decimated the population. If the plague had occurred in earlier Saxon times when the land was cultivated by individual settlers, many fields would certainly have been abandoned for lack of labor and therefore would have returned to the original forested state. The feudal system, however, created a different social situation. As one shepherd with a flock of sheep could deal with much more land than many plowmen and oxteams, the lords could keep their land in productive use in spite of labor shortage. The large flock of sheep that were thus maintained during the fourteenth century destroyed much of the new tree growth and converted the land into pasture.[17]

Traditional sheep-grazing, unlike the munching of cattle, cropped the grass to a lawn-like texture, which stimulated the growth of the grass and of many wildflowers—rock rose, wild thyme, scabious, etc.—which make the Sussex Downs smell, in the words of Rudyard Kipling "like dawn in paradise." The multitude of insects, and especially the butterflies, also depend on these sheep-grazed plants. Thus in this case again, the creation and maintenance of a highly desirable ecosystem is dependent on a multiplicity of random factors which result in the inhibition of tree growth by the animal population.[18]

Other desirable ecosystems, in contrast, have emerged from the proliferation of certain trees. In prehistoric time, the Adriatic Coast, from the Po Valley to south of Ravenna, was covered with oak and beech. During the fourth and fifth centuries the monks introduced *Pinus pinea* for its nuts and for esthetic reasons. The pines thrived, escaped, and established themselves on the hills. Thus was created the harmonious mixture of broadleaf trees and pines that Dante described in *Purgatory*.[19]

Rainfall, wind, and drought have, of course, exerted an enormous influence on the natural evolution of soils—for example, in the "dust bowl" of the southern Great Plains, which is in the region of the so-called "Brown Soils." It had been feared that the agricultural value of the land would be destroyed by the tremendous dust storms of the 1930s, but in fact the dust bowl has produced bumper crops during the past two decades, in part as a result of wiser agricultural practices, but also because of somewhat greater rainfall. As mentioned earlier, dust storms have occurred repeatedly under natural conditions in the past and they certainly will occur again in the future. Rainfall is one of those random events that has an unpredictable influence on the evolution of the dust bowl ecosystem.

Other factors in the evolution of natural ecosystems are the changes that have occurred in the chemical composition of agricultural soils, chiefly but not exclusively as a consequence of agricultural use. For example, the organic content of prairie soils has decreased since the land has been put under cultivation. In contrast, the people of Western Europe learned long ago that most of their soils (of the Gray-Brown Podzolic types) have to be manured for satisfactory crop yield. Similarly, the early colonists of New England learned from the Indians to put a fish into each hill of corn, the only fertilizer the Indians seem to have known. The evidence seems clear that the Gray-Brown Podzolic soils in Europe, and perhaps in some parts of the United States, are now richer in organic matter and more fertile than they were in the original forest wilderness. Agricultural practices and the general policies of land management may now play a role as important as that of natural forces in the evolution of ecosystems.[20]

THE DELIBERATE CREATION OF ARTIFICIAL ECOSYSTEMS

Industrial civilization has profoundly transformed the surface of the Earth in many areas, but not as extensively as commonly believed. It is certain, in any case, that most landscapes received a more dramatic shaping by human hands during the Neolithic and Bronze cultures than at any time after, including our own times. Surprising as it may seem, the appearance of the Earth as we know it today is largely the creation of people who transformed it with bare hands, simple tools, and a few domestic animals, several thousand years ago.

In many parts of the world which have been occupied for long periods of time, human beings have deliberately created artificial ecosystems we can regard as humanized analogues of natural communities. In Asia and Europe, particularly, large areas have been patiently shaped through a trial-and-error process which is similar in some ways to organic evolution. Early settlers have thus created artificial ecosystems possessing high levels of ecological diversity and stability in areas where the original ecosystem of the wilderness has been completely destroyed.

Ecological diversity has been increased, of course, by the deliberate or accidental importation of animal and plant species. As already mentioned, this process began during the Stone Age and it has continued ever since on a larger and larger scale. Hawaii provides a striking example of such an increase in ecological diversity through the introduction of foreign species.[21] Before the white settlements, the Hawaiian biota was relatively simple and disharmonic. It was deficient in terrestrial vertebrates; it had no pine trees, no oaks, no maples, no willows, no fig trees, no mangroves, one single species of palm, and only a few insignificant orchids. Needless to say, countless species have now been introduced into Hawaii and other Pacific islands and have prospered there even better than they did in their native habitats—as has been the case for many species transferred from one part of the world to another.

Many plant species imported from other continents have indeed established ecological communities which are stable even though of foreign origin, as has the Australian eucalyptus in North Africa and in California. The olive tree was first introduced from East Asia during the Stone Age and is now one of the most characteristic elements of Mediterranean vegetation; a large and beautiful olive grove near the Delphi site is at least several thousand years old, as shown by its mention in the Homeric writings.

Even more interesting than the introduction of plants and animals from other parts of the world, however, is the fact that an increase in ecological diversity has resulted from deliberate changes in the environment. Human analogues of natural communities have emerged from the replacement of the wilderness by a multiplicity of new micro-habitats characterized by a kind of environmental diversity different from that of the original environment, which, as was the case for certain semi-desertic and forested areas, may have been somewhat monotonous.

There is such a wealth of successful artificial ecosystems throughout the world that it will suffice to mention here a few types, selected because they correspond to sharply different climatic regions.

In much of the Arab world, natural ecosystems have been almost completely replaced by artificial ones; although many of them have undergone degradation, others have been more successful. It is largely through human intervention that the Nile Valley has long been profoundly different from the neighboring desert. The Ghuta orchards of Damascus, the palm groves of Matmata in Tunis, and the artificial oases of the Maghreb are examples of productive and pleasant ecosystems created and maintained by human effort under arid conditions. Thanks to the creation of ecosystems designed to make use of every drop of water, the desertic areas of Israel now produce large yields of food as well as medicinal plants, even under the extremely dry conditions that prevailed during the 1962-63 season. Productive farms operate even in the arid wastelands of the Negev, using only the few millimeters of rain which constitute the total annual precipitation in that area.[22]

The wet-rice economy of Southern China is based on an artificial ecosystem as remarkable for its complexity and stability as for its productivity. The rice paddies are usually located in valleys that can be subjected to controlled flooding. The water may be used first to irrigate vegetable gardens, where it accumulates organic matter; it is then able to produce a fine bloom of algae which in turn supports various forms of animal life. Oysters and other shellfish are raised in it, and it sustains shoals of finfish. The ecosystem is kept in smooth operation by natural processes of fertilization and pest control; chemical pesticides are not used. Pigs, chickens, ducks, and frogs feed on the refuse and become themselves succulent parts of the human diet. Weeds that are missed in the course of this cleaning process are pulled up and used as mulch or as feed for the pigs and fish. All nutrients are recycled into the system, including animal and human dung.

Each animal species of the wet-rice ecosystem has its own niche. Some species of carp are surface eaters, others grass eaters, still others are plankton feeders for different water levels; the mullet acts as a bottom detritus feeder. This artificial ecosystem is probably as stable as any in the world and as productive of food with regard to both quantity and diversity.

Another completely artificial ecosystem is that of the hedgerows which line up the country roads in England and France and also separate the fields; most of them were created during the very early Middle Ages and perhaps earlier. On the other hand, the patchwork of small fields and hedges which is commonly regarded as the typical landscape of East Anglia was laid down much later; it was created by law in the eighteenth and nineteenth centuries by the Inclosure Acts.

Whether of ancient origin or dating from Inclosure times, the hedgerows now constitute an ecosystem of great diversity and charm. Unlike American hedges, they are not rows of trimmed shrubs of a single species, but complex populations of trees, shrubs, flowering plants, grasses, small mammals, song birds, and a host of invertebrates. They serve as reservoirs for animal and plant species that probably would not do well either in the primeval forest or in a completely cleared landscape. In other words, they have an ecological diversity of their own. Hedges contribute in many other ways to the quality of the landscape, for example by providing habitats for spiders and other natural enemies of insects that prey on crops; by acting as windbreaks and thus protecting the land; by providing shade for domestic animals and hikers.[23]

Artificial ecosystems have thus been created all over the world, and it is certain that new ones will continue to be created. In the course of history, East Anglia passed from the stage of the primeval forest to the stage of the opened Celtic and Saxon fields, then to the stage of the small enclosures. Further changes are now in progress because the small fields created by the Inclosure Acts are no longer economically viable since they do not lend themselves to the use of modern large-scale agricultural equipment. The movement back to the large open fields of the Celtic-Saxon type will certainly disturb the ecological equilibrium that emerged in the enclosures, but with proper management it probably can lead to other kinds of successful ecosystems.

There are many examples, indeed, of new types of artificial ecosystems emerging in other types of environments. The Lake St. John region of Quebec is characterized by sand plains and granitic outcrops overlain with shallow soil; its natural vegetation consists chiefly of inferior tree species such as the jack pine.[24] Controlled periodic burning, however, keeps the forest down and allows the spontaneous growth of blueberry, which finds successful outlets in the Montreal and New York markets. In other sites, the jack pines or birch trees can be replaced by aspens, trees that provide a kind of wood highly suitable for the manufacture of plastics.

The mismanagement of tropical forests is presently one of the most dangerous threats to the Earth's ecosphere.

However, there are possibilities of solutions even in this case because with better understanding of tropical ecology, and especially with awareness of the fact that the ground should never be left bare under tropical conditions, a system of perennial crops may be developed as the mainstay of successful farming in artificial ecosystems derived from the rainforest.[25]

One can thus almost take it for granted that various "economies de rechange" will progressively emerge from ecological knowledge and will continue the creation of artificial ecosystems which has been a universal accompaniment of human life for many millenia.

Ecosystems have continuously evolved in the course of time, first through the influence of random natural events, and now increasingly because of human interventions. Whatever the mechanism responsible for the change, the ecosystem can persist only if it is in energy balance, but the level of energy input and output can be profoundly modified by natural and human forces. The utilization of solar energy is, for example, one of the factors which affects the level of energy balance.

In natural ecosystems, even in the tropics, much less than 0.1 percent of the solar energy impinging on a given area is trapped by the vegetation. The percentage may reach close to 0.3 in certain types of agriculture, and has been brought up to 4 under special experimental conditions. In practice, the most effective use of solar energy by advanced agriculture has been achieved with corn, sorghum, sugar cane, and a few other tropical species. With knowledge and experience, the mutualism between humankind and the Earth can result in artificial ecosystems increasingly creative of energy and resources.

CONCLUSIONS

The quality of landscapes and waterscapes can often be restored rapidly, by fairly simple measures, even when environmental degradation has gone far and has been long lasting.

Natural ecosystems can be transformed into humanized ecosystems which are biologically diversified, ecologically stable, and productive of new human values.

PRESIDIO PARK, SAN FRANCISCO

The history of San Francisco's Presidio Park, one of America's most beloved landscapes, will suffice to illustrate that human interventions into Nature can be creative with lastingly beneficial effects.

At the present time, the extensive wooded lands of the Presidio contrast sharply with the bareness of the surrounding areas at the northern tip of the San Francisco peninsula. This contrast is not a freak of Nature, but the result of a detailed "Plan for the Cultivation of Trees upon the Presidio Reservation" presented on 23 March 1883, by Major W.A. Jones, to the departmental commander Major General Irwin McDowell.

Some fifty-five thousand acacia pine, cypress and eucalyptus trees and five thousand native redwood, spruce and madrone trees were planted. For some period of time, at the invitation of the Army, native San Franciscans joined to participate in Arbor Day celebrations and help reforest the Presidio. The actual number of trees planted is not known, but the planting continues year by year. In addition, there is a very large amount of natural seeding, especially pine and cypress. Of the Presidio's 1400 acres today approximately 280 acres are defined as woodland.

No commercial use is made of any of the trees of the Presidio. The only use made of the wood is for firewood though the source is restricted to "leftovers" from topping, thinning and a few trees that must be removed. At the present time, the spirit of conservation and preservation is so strong that it is difficult to get permission to remove or thin trees even in places where it is absolutely necessary. The only major removal of trees from the Presidio has been by the state when the Golden Gate Bridge was built.

PRESIDIO FROM STREET CAR STATION - 1882. The Presidio when Major General Irwin McDowell was Commanding General of the Post. Trees in the foreground were planted at his direction in order to protect as well as beautify the old post which has been buffeted by wind and storms for 106 years.

PRESIDIO OF SAN FRANCISCO–1887. This photograph taken after the death of Major General Irwin McDowell shows the first evidence of the extensive woodland groves of today's Presidio recommended under the "Jones Plan."

This photograph shows the view from the south entrance of the Golden Gate Bridge looking towards the Presidio. The separate grove of trees in the foreground was contiguous until severed by the bridge approach and toll plaza.

A photo looking north along Baker's Beach towards Golden Gate Bridge on Fort Scott.

ENVOI

Nature is like a great river of materials and forces that can be directed in this or that channel by human interventions. Such interventions are often needed because the natural channels are not necessarily the most desirable, either for the human species or for the Earth. Nature often creates ecosystems which are inefficient, wasteful and destructive. By using reason and knowledge, human beings can manipulate the raw stuff of nature and shape it into ecosystems that have qualities not found in the wilderness. They can give a fuller expression to many potentialities of the Earth by entering with it in a relationship of symbiotic mutualism.

The Earth is neither an ecosystem to be preserved unchanged, nor a quarry to be exploited for selfish and short-range economic reasons, but a garden to be cultivated for the development of its own potentialities and the potentialities of the human species. The goal of this relationship is not the maintenance of the status quo but the emergence of new phenomena and new values. Millenia of experience show that, by entering into a mutualistic symbiosis with the Earth, humankind can invent and generate futures not predictable from the deterministic order of things, and thus can engage in a continuous process of creation.

REFERENCE NOTES

1. Aldo Leopold, *A Sand County Almanac and Sketches Here and There* (New York: Oxford University Press, 1949), p. 224.

2. J. Cairns, Jr., K.L. Dickson, and E.E. Herricks, ed., *Recovery and Restoration of Damaged Ecosystems* (Charlottesville: University Press of Virginia, 1977); C.S. Holling, "Resilience and Stability of Ecological Systems," *Annual Review of Ecology and Systematics* 4 (1973): 1-23.

3. John J. Kupa and William R. Whitman, *Land-cover Types of Rhode Island: an ecological inventory*, University of Rhode Island Agricultural Experiment Station Bulletin 409 (Kingston, 1972).

4. Rutherford H. Platt, "The Loss of Farmland: Evolution of Public Response," *Geographical Review* 67 (1977): 93-101; Paul E. Waggoner and George R. Stephens, "Return of the Forest," *Natural History* 82 (1973): 82-83.

5. Lee W. DeCraff, "Return of the Wild Turkey," *The Conservationist* (October-November, 1973): 24-27.

6. Lars Emmelin, "The Beaver–Conservation Problems," *Current Sweden*, Environment Planning and Conservation 61 (February, 1976). Available from Swedish Information Service, New York.

7. Erik P. Eckholm, *Losing Ground: Environmental Stress and World Food Prospects* (New York: Norton, 1976); Erik P. Eckholm, "The Deterioration of Mountain Environments," *Science* 189 (1975): 764-70; David Pimentel *et al.*, "Land Degradation: Effects on Food and Energy Resources," *Science* 194 (1976): 149-55.

8. Anthony Smith, *The Seasons–Life and its Rhythms* (New York: Harcourt Brace, 1970), p. 162.

9. Sturla Fridriksson, *Surtsey: Evolution of Life on a Volcanic Island* (London: Butterworth, 1975).

10. Samir I. Ghabbour, "National Parks in Arab Countries," *Environmental Conservation* 2 (1975): 45-46.

11. Nicholas Wade, "Sahelian Drought: No Victory for Western Aid," *Science* 185 (1974): 234-37.

12. Personal information.

13. Cairns, Dickson and Herricks, *Recovery and Restoration*, pp. 24-134.

14. David J. Parsons, "The Role of Fire in Natural Communities: An Example from the Southern Sierra Nevada, California," *Environmental Conservation* 3 (1976): 91-99; H.E. Wright, Jr., "Landscape Development, Forest Fires, and Wilderness Management," *Science* 186 (1974): 487-97.

15. D.F. Costello, *The Prairie World* (New York: Crowell, 1969).

16. David W. Ehrenfeld, "Man Intervention in Living Systems," Proceedings of *Engineering for the Environment*, held May 6-7, 1971, in Wakefield, Mass., sponsored by the University of Massachusetts School of Engineering.

17. Brenda Colvin, *Land and Landscape* (London: John Murray, 1948), p. 24.

18. Jon Tinker, "Farming and Conservation," *New Scientist* 62, (1974): 219.

19. Nicholas Mirov, "The Pines of Ravenna," *Natural History* 80 (1971): 24-26.

20. Roy Simonson, "The Soil Under Natural and Cultural Environments," *Journal of Soil and Water Conservation* 6 (1951).

21. F.R. Fosberg, ed., *Man's Place in the Island Ecosystem* (Honolulu: Bishop Museum Press, 1965), p. 61.

22. Michael Evanari, Leslie Shanan, and Naphtali Tadmor, *The Negev: Challenge of a Desert* (Cambridge: Harvard University Press, 1971); Bernard Mandelbaum, ed., *Assignment in Israel* (New York: Harper & Brothers, 1960), p. 174.

23. David W. Ehrenfeld, *Conserving Life on Earth* (New York: Oxford University Press, 1972), p. 276; Jon Tinker, "The End of the English Landscape," *New Scientist* 64 (1974): 722-27.

24. Pierre Dansereau, "Ecological Impact and Human Ecology," in F. Fraser Darling and John P. Milton, eds., *Future Environments of North America* (Garden City, New York: The Natural History Press, 1966), 425-62.

25. Edward G. Farnworth and Frank B. Golley, eds., *Fragile Ecosystems: Evaluation of Research and Applications in the Neotropics* (New York: Springer-Verlag, 1974).

ABOUT THE AUTHOR

René Jules Dubos is an internationally known and respected scientist and author. He came to the United States from his native France to attend graduate school at Rutgers University, where he received a doctoral degree in microbiology. He remained in this country to become a faculty member at Rockefeller University, from 1927 (with the status of Fellow) until his retirement in 1971 as emeritus professor of pathology.

Dr. Dubos is a member of the National Academy of Sciences and the winner of a dozen awards in the United States and Europe for the excellence of his scientific work. His concern with the effects of environmental forces on human life has been lifelong, and has intensified in the past two decades. He was one of the very few members of the National Academy to defend Rachel Carson when her book *Silent Spring* was under criticism. This concern led him to write several books on the human environment, one of which—*So Human an Animal*—was awarded the Pulitzer Prize (1969); another—*Man Adapting*—is one of the leading environmental books of the 1970s.

REUBEN GUSTAVSON

Reuben G. Gustavson, born in Denver on April 6, 1892, was the son of an immigrant carpenter. Because of a physical ailment caused by a boyhood accident, his father decided that Reuben could never make a living as a carpenter and insisted that he take the commercial course in high school. Reuben's job was as an office worker for a railroad. The likelihood of an academic career for him was extremely slight—particularly one that might lead to national and international repute.

Volunteer work in the research laboratory of a tuberculosis sanitorium fired Gustavson's desire to study for a scientific career. Lack of money (as usual) and admission (language) requirements seemed insurmountable obstacles. A perceptive admissions officer found a way around the language requirement by accepting an examination in Swedish (the language of the Gustavson home) in lieu of a high school language course. Had it not been for this wise judgment, our country might well have been deprived of Reuben Gustavson's enormous contribution to science and education.

He was to be a teacher of chemistry at Colorado State University (1917-20) and the University of Arizona (1959-67); teacher and researcher in chemistry and chairman of the department at the University of Denver (1920-37) and the University of Colorado (1942-43); vice president and dean of faculties at the University of Chicago (1945-46); president of the University of Colorado (1943-45), and chancellor of the University of Nebraska (1946-53). He became president of the Resources for the Future (RFF) foundation during its formative years (1953-59) and chairman of its board of directors (1961-71). He died in 1974.

Highlights of Gustavson's academic career include early and effective actions in the struggle for equal rights for minorities and in strengthening academic freedoms. He was less successful in efforts to eliminate professionalism in intercollegiate athletics. He is noted for pioneering research in sex hormones, specifically estrogen, as a means of population control. Most notable was his ability as a teacher. Even after his retirement at age sixty-seven he taught a TV course in beginning chemistry for non-science majors at the University of Arizona so successfully that it became the sole freshman course in the subject.

In recognition of these achievements, the RFF offered matching grants to four universities to establish this memorial lecture series.